RAILWAYS IN NORTH AND MID WALES
IN THE LATE 20TH CENTURY

RAILWAYS IN NORTH AND MID WALES

IN THE LATE 20TH CENTURY

PETER J. GREEN

PEN & SWORD TRANSPORT

AN IMPRINT OF PEN & SWORD BOOKS LTD.
YORKSHIRE – PHILADELPHIA

First published in Great Britain in 2022 by
Pen and Sword Transport
An imprint of
Pen & Sword Books Ltd
Yorkshire - Philadelphia

Copyright © Peter J. Green, 2022

ISBN 978 1 39909 122 0

The right of Peter J. Green to be identified as the author of this work has been asserted by him in accordance with the Copyright, Designs and Patents Act 1988.

A CIP catalogue record for this book is available from the British Library.

All rights reserved. No part of this book may be reproduced or transmitted in any form or by any means, electronic or mechanical including photocopying, recording or by any information storage and retrieval system, without permission from the Publisher in writing.

Typeset by SJmagic DESIGN SERVICES, India.

Printed and bound in India by Replika Press Pvt. Ltd.

Pen & Sword Books Ltd incorporates the Imprints of Pen & Sword Books Archaeology, Atlas, Aviation, Battleground, Discovery, Family History, History, Maritime, Military, Naval, Politics, Railways, Select, Transport, True Crime, Fiction, Frontline Books, Leo Cooper, Praetorian Press, Seaforth Publishing, Wharncliffe and White Owl.

For a complete list of Pen & Sword titles please contact

PEN & SWORD BOOKS LIMITED
47 Church Street, Barnsley, South Yorkshire, S70 2AS, England
E-mail: enquiries@pen-and-sword.co.uk
Website: www.pen-and-sword.co.uk

or

PEN AND SWORD BOOKS
1950 Lawrence Rd, Havertown, PA 19083, USA
E-mail: Uspen-and-sword@casematepublishers.com
Website: www.penandswordbooks.com

CONTENTS

Introduction 6

Acknowledgements 8

Map of the Passenger Lines of North and Mid Wales 9

The North Wales Coast at its best ....Photo 1

The North Wales Coast Line, Chester to Llandudno ....Photos 2–60

Llandudno Junction to Trawsfynydd ....Photos 61–77

Llandudno Junction to Gaerwen Junction ....Photos 78–98

The Amlwch Line ....Photos 99–102

Gaerwen Junction to Holyhead ....Photos 103–114

Chester to Wrexham ....Photos 115–121

Wrexham to Shotton and Hawarden Bridge ....Photos 122–129

Wrexham to Chirk ....Photos 130–138

The Llangollen Railway ....Photos 139–140

The Welshpool and Llanfair Light Railway ....Photos 141–142

The Cambrian from Welshpool to Pwllheli ....Photos 143–206

Dovey Junction to Aberystwyth ....Photos 207–218

The Vale of Rheidol Railway ....Photos 219–223

Steam-Hauled Special Trains ....Photos 224–231

Bibliography 139

Index 140

INTRODUCTION

During the years that I have spent photographing railways, both at home and abroad, there have been various experiences that I remember very clearly. As an enthusiast with a particular liking for Class 37 diesels, a number of these experiences relate to the railways in Wales where, at different times, the class has seen regular use on most of the lines. My memories include waiting at Barmouth Bridge for the special train marking the reopening of the bridge to locomotive-hauled trains in April 1986, Class 37 diesels with 'Cambrian Coast Express' headboards during the same year, and locomotive-hauled trains running along the North Wales coast at locations such as Llanfairfechan and Penmaenmawr. Photographs of all of these appear in these pages.

In this book, we look at the British Railways lines and the trains that ran on them in the northern part of Wales in the years between 1980 and 2000, although a few photographs from earlier years are also included to help to complete the picture. In my opinion at least, the coastal and mountain scenery around the railway lines in North and Mid Wales is among the best in Great Britain. The railways that feature here are the North Wales Coast line and its branches, between Chester and Holyhead, the Cambrian line from Welshpool to Aberystwyth and Pwllheli, and the Welsh section of the Shrewsbury to Chester line. The Welshpool and Llanfair Light Railway, closed by British Railways in 1956 and reopened as a heritage railway, and the Llangollen Railway on a section of the former Ruabon to Barmouth line also feature, as does the Vale of Rheidol Railway, sold into private ownership in 1989. I also thought that the book would not be complete without a few photographs of the steam specials that regularly ran on the main lines.

Researching the captions for the photographs in this book has brought home to me just how much of the railway infrastructure that I recorded has disappeared. Stations have been modernised and, in particular, the signal boxes and the associated mechanical signalling largely belong to the past. The motive power and the nature of the trains has also changed significantly over the years and, while all of this is beneficial to the traveller, it has also made the railway more standardised and less interesting to many enthusiasts.

One of the advantages of my home town of Worcester is that its location makes day trips possible to almost anywhere in England and Wales. The disadvantage is that, with many destinations to choose from, I tended not to concentrate too much on any particular area. Nevertheless, I made many visits to Wales in the 1980s and 1990s, although I regret not spending more time in North Wales before the Class 37s took over many of the passenger trains along the coast.

I also visited South Wales many times, and it is my intention that photographs from those trips will be included in a second volume.

Nowadays, most of my visits to Wales are to view, photograph, and travel on the many excellent narrow gauge railways, including the Vale of Rheidol and the Welshpool and Llanfair.

Peter J. Green
Worcester, England,
2021.

ACKNOWLEDGEMENTS

Paul Dorney has provided me with a number of photographs for use here and, as well as discussing various aspects of this book with me, he has always had answers to the many questions I have asked.

Steve Turner has kindly supplied photos of the Amlwch and Penyffordd lines, and the nuclear flask trains between Blaenau Festiniog and Trawsfynydd. I started visiting the North Wales Coast line when the 'Peaks' and Class 40s had largely been displaced by other motive power. Steve has also helped with photos of these.

James Billingham has once again assisted me with some of the captions. Many of the photographs included here were taken on visits to Wales that we made together.

Val Brown has again checked my text and corrected the many mistakes I have made.

My thanks go to you all. This book would be poorer without your help.

MAP OF THE PASSENGER LINES OF NORTH AND MID WALES • 9

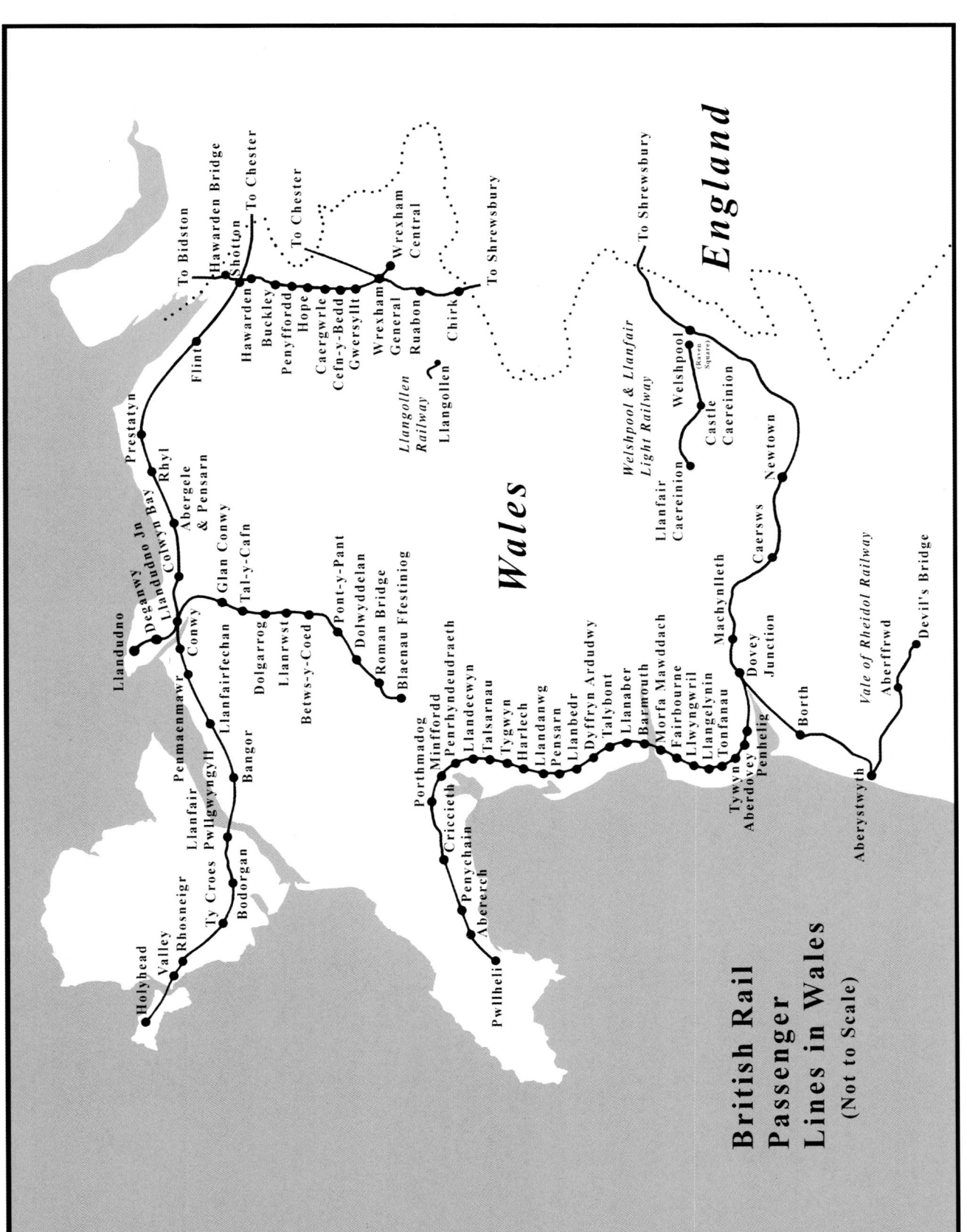

British Rail Passenger Lines in Wales (Not to Scale)

This map shows British Rail's passenger routes during the period covered by this book. Also included is the preserved Llangollen Railway, which runs on a section of the former Ruabon to Barmouth line, closed in 1965, and the narrow gauge Welshpool and Llanfair Light Railway, closed by British Railways in 1956 and reopened as a heritage railway in 1963. British Railways' narrow gauge Vale of Rheidol Railway was sold into private ownership as an operating railway in 1989. Other narrow gauge railways which run on old British Railways trackbeds are not included.

THE NORTH WALES COAST AT ITS BEST

1. With a blue sky and the sun shining on Conwy Bay, English Electric Type 3 37509, in Transrail livery, passes Llanfairfechan with the 10.23 Bangor to Crewe. Transrail was a trainload rail freight operator based in St Blazey. 24 June 1995.

THE NORTH WALES COAST LINE, CHESTER TO LLANDUDNO

2. Trains heading along the North Wales Coast line pass through Chester, a major railway junction in England close to the border with Wales. Before resignalling, English Electric Type 4 40024 passes Chester No. 2 signal box, closed in 1984, as it heads a Speedlink feeder service to Warrington with aluminium from Valley, Anglesey. A Sulzer Type 2 waits with a freight in the yard next to the station. 24 February 1984. (*Paul Dorney Photo*)

3. A diesel multiple unit (DMU), forming the 11.44 Llandudno to Nottingham, heads east from Sandycroft. Three-car Class 116 set T323 is leading. The Class 116 DMUs were built by British Rail Derby between 1957 and 1961. 31 August 1991.

4. Sandycroft station, opened in 1884 by the London and North Western Railway (LNWR), was closed in 1961. A loaded coal train from Point of Ayr Colliery to Fiddlers Ferry Power Station passes Sandycroft signal box, closed and demolished in 2005, behind 20013 and 20090. 21 September 1991.

5. Sulzer Type 4 'Peak' 45140 approaches the site of Saltney Ferry station at Mold Junction with the 1M58 08.20 Newcastle to Llandudno. The station, close to the English border, was opened in 1891 by the LNWR on their line from Chester to Holyhead. It was closed in 1962. The line to Mold and on to Denbigh left the North Wales Coast line here. Mold Junction No. 1 signal box, closed in 2005, is on the right. 30 January 1987. (*Steve Turner Photo*)

6. Four and a half years later, much of the track at Mold Junction has been removed. For many years, a locomotive depot and the main freight marshalling yard on the line was located here. A Trainload Coal Day was held on the line in August 1991 when Trainload Coal locomotives were used on various passenger services. Here, 58003 *Markham Colliery* and 58007 *Drakelow Power Station* pass Mold Junction with the 1T58 15.40 Crewe to Llandudno. 11 August 1991.

7. 'Peak' 45139 approaches Mold Junction with the 10.44 Holyhead to Newcastle. The old marshalling yard can be seen on the left with Hawarden Airport beyond. The line to Mold and Denbigh curved away to the left of the sidings. 30 January 1987. (*Steve Turner Photo*)

8. Appropriately on Trainload Coal Day, a loaded coal train, from Point of Ayr Colliery to Fiddlers Ferry Power Station at Warrington, passes Mold Junction behind English Electric Type 1 locomotives 20143 and 20214. On the left, the tracks of the marshalling yard at Mold Junction have been removed. 11 August 1991.

9. Flint station building, designed by Francis Thompson, was opened in 1848. Sprinter 156 403, on the left, is working the 07.01 Stalybridge to Bangor, while 156 460, forming the 08.17 Llandudno to Stalybridge, waits in the platform. The signal box, built to a London Midland and Scottish Railway (LMSR) design, closed in 1989. 22 July 1989.

10. With the derelict signal box on the left, 40008 passes Bagillt with the 1D57 13.45 Manchester Victoria to Bangor. The unofficial headboard on the locomotive reads 'The Welsh Whistler'. The Class 40s were called 'Whistlers' by enthusiasts because of the whistling noise made by their turbochargers. The station was closed in 1966 and the signal box in 1973. 16 August 1982. (*Steve Turner Photo*)

11. In the same location as the previous photo but taken thirteen years later, the station platforms are even more overgrown and the old signal box has been removed. Bagillt goods shed still stands and can be seen in private use, on the right. English Electric Type 3 37425 *Sir Robert McAlpine/Concrete Bob* passes with the 1D76 15.24 Crewe to Holyhead. 24 June 1995.

12. English Electric 37429 *Eisteddfod Genedlaethol,* in Regional Railways livery, approaches the closed station at Bagillt with the 14.55 Holyhead to Crewe. Note the sluice gate on the right, used to control water flow near the coast. 8 July 2000.

13. Pictured between Holywell Junction and Bagillt, English Electric Class 37/4 37415 *Mt Etna*, in InterCity livery, heads empty fuel tanks from Holyhead depot to the Shell refinery at Stanlow. 1 July 1995.

14. With the bridge at Holywell Junction in the distance, 31134, in Civil Engineers' yellow and grey livery, heads the 1K69 13.56 Holyhead to Crewe towards Bagillt. 24 June 1995.

15. An eastbound freightliner from Holyhead passes Holywell Junction with Brush Type 4 47245 in charge. Until 1991, when the service was withdrawn, container ships crossed the Irish Sea from Holyhead to Dublin and Belfast. 22 July 1989.

16. Holywell station was opened in 1848. In 1912, a branch line to Holywell Town, in the centre of Holywell, was opened to the east of the station, which was then renamed Holywell Junction. The signal box opened in 1902, replacing an earlier wooden box. Here, Class 37/4 37414 *Cathays C&W Works 1846-1993*, in Regional Railways livery, passes Holywell Junction station with the 1J44 13.30 Holyhead to Manchester Victoria. The station closed in 1966 and the signal box in 2018. 28 August 1993.

17. From the railway activity, it would appear that permanent way work is scheduled in the area. Former Glasgow Eastfield Class 37/4 37408 *Loch Rannoch,* in Large Logo livery, passes Holywell Junction with an eastbound ballast train while a second ballast train stands in the loop. 27 September 1997.

18. Class 150 Sprinter 150 101 heads for Llandudno near Mostyn. The ship in the background is the TSS *Duke of Lancaster*, built by Harland & Wolff for the British Transport Commission in 1956 for use on the Heysham to Belfast service. It served until 1978 when it was sold to a Liverpool firm which operated it as an entertainment site, the 'Mostyn Funship'. It is beached at Llanerch-y-Mor, between Holywell Junction and Mostyn. 31 August 1991.

19. Birmingham Railway Carriage and Wagon Company (BRCW) Type 3 33003 approaches Mostyn with the 1D27 11.15 Crewe to Bangor. 26 March 1986. (*Steve Turner Photo*)

20. English Electric Type 4 40015 passes Mostyn with the 1K28 09.20 Holyhead to Crewe. The Class 40 was named *Aquitania* until December 1973. 14 August 1982. (*Steve Turner Photo*)

21. Fitted with an unofficial *Aureol* nameplate and carrying its number on its nose, English Electric Class 40 40012 leads the 1J53 15.17 Holyhead to Manchester Victoria past Mostyn. The name *Aureol* was officially removed in 1973. The locomotive is now preserved at the Midland Railway, Butterley. 14 August 1982. (*Steve Turner Photo*)

22. 'Peak' 45125 approaches the old station site at Mostyn with the 1M63 11.00 Scarborough to Bangor. Mostyn Docks are out of sight, to the left of the photograph. The locomotive is now preserved at the Great Central Railway, where it carries its original number, D123. 15 August 1984. (*Paul Dorney Photo*)

23. Type 4 40184 heads the 1J22 13.58 Bangor to Manchester Victoria past Mostyn signal box. Formerly Mostyn No. 1 signal box, it is a modified LNWR Type 4 design, built in 1902. The box was closed in 2017 but remains as a Grade II listed building. 14 August 1982. (*Steve Turner Photo*)

24. Viewed from the footbridge next to the signal box, a Metro-Cammell Class 101 DMU passes the site of Mostyn station as it heads for Manchester Victoria. The station, opened in 1848, was closed in 1966. The station building and goods shed remain, out of sight on the left. 2 June 1985.

25. Running alongside the River Dee estuary near Ffynnongroyw, 40009 heads the 1E82 13.53 Llandudno to York towards Mostyn. Point of Ayr can be seen in the distance. 14 August 1982. (*Steve Turner Photo*)

26. With Point of Ayr Colliery in the background, two-car Metro-Cammell Class 101 DMU set LO603 passes Talacre. The working is the 16.13 Crewe to Holyhead. Before closure in 1996, frequent coal trains ran along the North Wales Coast between Point of Ayr Colliery and Fiddlers Ferry Power Station at Warrington. 31 August 1991.

27. Class 46 'Peak' D172 *Ixion* and Brush Type 4 47834 *Fire Fly* head east at Talacre with a test train from Crewe Works. The 'Peak' has been painted in British Railways green livery for Pete Waterman. D172 was renumbered 46035 and later became 97403, when it was used by Derby Research Centre. It is preserved at Peak Rail, Rowsley. Talacre signal box, opened in 1903, is a LNWR Type 4 design. The box closed in 2018 and was sold for use as an office. Talacre station, opened in 1903, was closed in 1966. The up platform can be seen beyond the bridge. 18 August 1994. (*Paul Dorney Photo*)

28. Brush Type 4 47822 *Pride of Shrewsbury* heads the 09.19 Holyhead to London Euston near Gronant, between Talacre and Prestatyn. A driving van trailer is behind the locomotive. The Class 47 will be removed and an electric locomotive attached to the rear of the train at Crewe for the journey to London. The whole train is in Virgin Trains livery. 8 July 2000.

29. Class 142 Pacer 142 014 and a Class 150 Sprinter are pictured departing from Prestatyn and heading for Crewe. The Pacer was based on a Leyland National bus body. Ninety-six units were built by British Rail at Derby Litchurch Lane between 1985 and 1987. 22 July 1989.

30. Prestatyn station was opened by the Chester and Holyhead Railway (CHR) in 1848. After the LNWR acquired the CHR, a branch line was opened to Dyserth, with the current station opening in 1897. Here, 37429 *Eisteddfod Genedlaethol* passes the original station building and goods shed as it arrives at Prestatyn with the 1D68 09.46 Birmingham International to Holyhead. 4 March 1995.

31. Class 150/2 Sprinter 150 227, forming the 13.07 Crewe to Llandudno, heads away from the camera as it departs from Prestatyn. The station building is the only survivor of the three prefabricated buildings built for the station by the LNWR in the 1890s. 22 July 1989.

32. Brush Type 4 47491 *Horwich Enterprise* passes under the fine signal gantry at the east end of the station, as it departs from Rhyl with the 1E93 17.03 Holyhead to York. 22 June 1985.

33. Passing the old carriage shed, 47555 *The Commonwealth Spirit* arrives at Rhyl with a Manchester Victoria to Llandudno service. 22 June 1985.

34. 'Crompton' 33063 arrives at Rhyl with the 1D27 10.45 Stafford to Bangor. The train ran via the Crewe Independent Lines due to modernisation work at Crewe. Rhyl No. 1 signal box, on the left, is a LNWR Type 4 design dating from 1900. The box closed in 2018. 22 June 1985.

35. Rhyl station, opened in 1848, was enlarged by the LNWR and a new layout was operated from 1900. A branch line to Denbigh was opened in 1858, finally closing in 1965. Here, 'Peak' 45123 *The Lancashire Fusilier* arrives at Rhyl with the 1E99 11.14 Bangor to Newcastle. 22 June 1985.

36. 'Peak' 45140 is about to depart from platform 2 at Rhyl with the 1M63 10.53 Scarborough to Holyhead. The spire of St Thomas' Church can be seen on the left. 18 July 1986. (*Steve Turner Photo*)

37. Pathfinder Tours' 'The Crompton Culminator' ran from Salisbury to Holyhead. Here, 'Cromptons' 33051 *Shakespeare Cliff* and 33030 run through the middle road at Rhyl as they head for Holyhead. A wreath is carried on the nose of 33051 because it is the day of Princess Diana's funeral. Rhyl No. 2 signal box, in the background, is a LNWR Type 4 design dating from 1900. It is similar to Rhyl No. 1 box but larger. Rhyl No. 2 box closed in 1990. 6 September 1997.

38. Sulzer Type 2 25078 runs under the signal gantry at the west end of Rhyl station as it heads east with a short freight. The train is probably bound for Mostyn Dock. 22 June 1985.

39. Class 37/4 37402 *Bont Y Bermo* approaches Abergele and Pensarn with the 1D39 10.15 Blackpool North to Holyhead. *Bont Y Bermo* is clearly not stopping, as can be seen from the down home signals for the main line and the platform loop. The locomotive received the name, previously carried by 37427, in 1994. 23 July 1994.

40. With Abergele and Pensarn station in the background, Brush Type 4 47852, in Virgin Trains livery, heads the 08.52 Holyhead to London Euston along the coast towards Rhyl. A driving van trailer is behind the locomotive. 15 April 2000.

41. English Electric Type 4 40135 heads a train of empty liquid chlorine tanks, from Associated Octel at Amlwch, through Abergele and Pensarn station. The first two grey tanks containing ethylene dibromide, used in the manufacture of anti-knocking agents for petrol, are bound for Octel's plant at Ellesmere Port. There is a guard's van and a barrier van at each end of the train. The Class 40 is now preserved at the East Lancashire Railway. 15 August 1984.

42. After calling at Abergele and Pensarn, a Metro-Cammell Class 101 DMU continues its journey to Manchester Victoria. The station was opened by the Chester and Holyhead Railway in 1848. The signal box is a LNWR Type 4 design, opened in 1902 and closed in 2018. 15 August 1984.

43. Brush Type 4 47239 takes the up middle road, since removed, through Abergele and Pensarn station with a ballast train from Penmaenmawr to St Helens. 15 August 1984.

44. In Railfreight red stripe livery, 47322 approaches Abergele and Pensarn with empty liquid chlorine tanks from Associated Octel at Amlwch. 24 July 1989.

45. In English, Welsh and Scottish Railway (EWS) maroon and gold livery, English Electric Type 3 37421 heads west from Abergele and Pensarn with the 10.07 Birmingham New Street to Holyhead. The station is behind the road bridge in the background. 5 March 2000.

46. In EWS livery, 37426 approaches Colwyn Bay with the 11.16 Bangor to Birmingham New Street. Colwyn Bay signal box, located at the west end of the station, was a British Railways (London Midland Region) Type 15 design. The box was opened in 1968 and closed in 1991. It was demolished approximately a week after this photo was taken. 11 March 2000.

47. Viewed from the Station Road bridge, the InterCity 125 unit forming the 08.58 Holyhead to London Euston, runs alongside the A55 North Wales Expressway near the site of Mochdre and Pabo station, closed in 1931, as it heads towards Colwyn Bay. The station site is now covered by the A55. 27 July 1996.

48. Viewed from the Conway Road bridge, 37420 *The Scottish Hosteller*, in InterCity livery, heads an eastbound loaded ballast train from Penmaenmawr away from Llandudno Junction. 23 February 1995. (*Paul Dorney Photo*)

49. English Electric Type 4 40012 approaches Llandudno Junction with Sunday's 1D40 14.36 Manchester Victoria to Llandudno. 15 August 1982. (*Steve Turner Photo*)

50. English Electric Type 3 37408 *Loch Rannoch* departs from the Llandudno Junction station stop with the 1K61 10.23 Bangor to Crewe. 3 August 1996.

51. The original Llandudno Junction station, located west of the current station, was opened in 1858. The larger present-day station was opened in 1897 and is the junction of the lines to Llandudno, Chester, Blaenau Ffestiniog and Holyhead. Here, Class 31/4 31432 departs from Llandudno Junction with the 13.50 Manchester Victoria to Llandudno. Class 156 Sprinter 156 455 stands in platform 2, on the left, and a Class 158 Sprinter and a Class 47, on a train of tanks for Associated Octel at Amlwch, are in the yard on the right. 17 July 1993.

52. Class 40 40012 heads the 1D40 14.36 Manchester Victoria to Llandudno away from Llandudno Junction station. Llandudno Junction signal box, formerly No. 2 box, was a LNWR Type 4 design. Opened in 1891, it was closed in 1985. 15 August 1982. (*Steve Turner Photo*)

53. 'Peak' 45109 rounds the curve on the approach to Llandudno Junction with the 1E53 13.24 Llandudno to Scarborough. The River Conwy estuary is on the left. 25 February 1984. (*Paul Dorney Photo*)

54. Deganwy station was opened by the LNWR in 1866. Here, 37422 *Cardiff Canton* runs through the station with the 1G81 12.06 Llandudno to Birmingham New Street. Deganwy signal box can be seen at the end of the down platform. It is a LNWR Type 5 design, dating from 1914. 21 August 1993.

55. Class 142 Pacer 142 015, forming a Llandudno to Bangor service, pauses at Deganwy. The station building has since been demolished. 23 July 1989.

56. In Regional Railways livery, 31455 *'Our Eli'* approaches Llandudno with the 1D61 07.35 Birmingham International to Llandudno. The course of Maesdu Golf Club is on the left. 7 August 1993.

57. Viewed from the Bryniau Road bridge, Class 31/4 31432 heads away from Llandudno with the 16.35 Llandudno to Liverpool Lime Street. 17 July 1993.

58. The signalman has quickly returned the signals to danger as Derby Lightweight Class 108 DMU set CH375, forming the 09.45 to Blaenau Ffestiniog, departs from Llandudno station. Llandudno signal box, built to a LNWR Type 4 design, was opened in 1891. 23 July 1989.

59. 'Crompton' 33015 reverses out of platform 3 at Llandudno station to run round its train, after arriving with the 1M84 12.00 from Cardiff Central. 8 July 1986. (*Steve Turner Photo*)

60. Llandudno station, opened in 1858, is three miles from Llandudno Junction on the North Wales Coast line. Rebuilding work in 2014 saw the demolition of part of the station to make room for a car park, a glazed concourse, and a bus interchange. Here, a Class 108 DMU, forming the 09.45 to Blaenau Ffestiniog, waits at platform 2 for departure time. The train is made up of M51938, M59137 and M52065 and the headboard ('Gwasanaeth Sul Dyffryn Conwy') indicates that it is a Conwy Valley Sunday service. 23 July 1989.

LLANDUDNO JUNCTION TO TRAWSFYNYDD

61. The line to Blaenau Ffestiniog and Trawsfynydd diverges from the North Wales Coast line just to the east of Llandudno Junction station. Here, 31421 *Wigan Pier* approaches the station with the 13.50 Manchester Victoria to Llandudno. The Blaenau Ffestiniog line curves away to the right. 21 August 1993.

62. Pathfinder Tours' 'The Trawsfynydd Trekker' ran from Bristol Temple Meads to Trawsfynydd. Here, the returning special train runs alongside the River Conwy, as it passes Glan Conwy behind 31238 and 31207. English Electric Type 1s 20187 *Sir Charles Wheatstone* and 20075 *Sir William Cooke* are on the rear of the train. 10 September 1994.

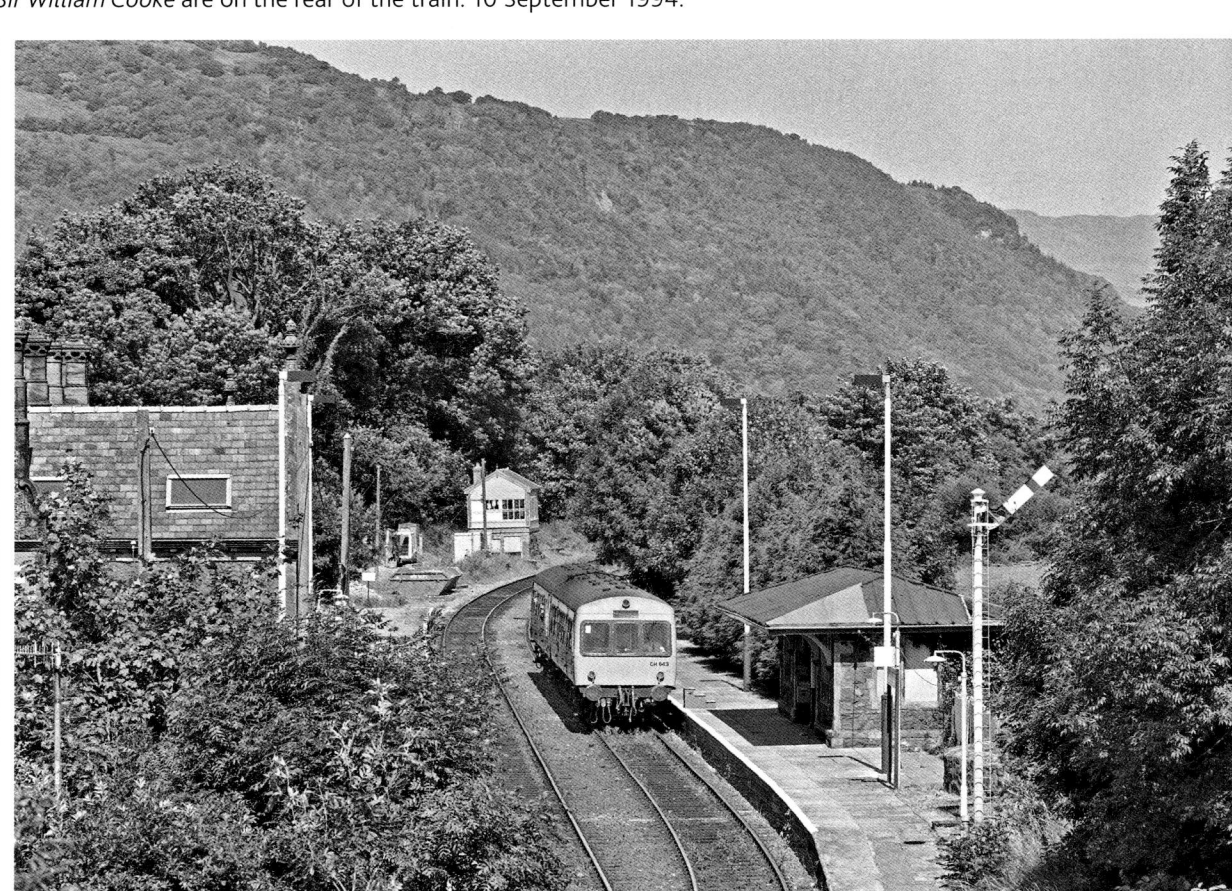

63. The original station at Llanrwst, opened in 1863, was resited by the LNWR in 1868. It was renamed North Llanrwst in 1989. Here, Class 101 DMU set CH643, forming a Llandudno to Blaenau Ffestiniog service, waits to depart from North Llanrwst. The signal box is a LNWR Type 4 design, dating from 1880. 14 July 1990.

64. Class 150 Sprinter 150 133, forming the 16.07 Llandudno to Blaenau Ffestiniog, departs from North Llanrwst. The station has the only passing loop on the Conwy Valley line between Llandudno Junction and Blaenau Ffestiniog. 27 September 1997.

65. Bettws-y-Coed station, opened in 1868, is fifteen miles from Llandudno Junction. Here, Class 101 DMU set L840 in Network SouthEast livery, forming a Llandudno to Blaenau Ffestiniog service, stands at Bettws-y-Coed. The miniature railway at the Conwy Valley Railway Museum is on the right. 6 February 1998. (*Paul Dorney Photo*)

66. DMU set CH647, forming the 11.00 Llandudno to Blaenau Ffestiniog service, is pictured near Dolwyddelan. A Class 101 unit is leading. 25 August 1990.

67. Soon after departing from Roman Bridge, DMU set CH647, forming the 12.15 Blaenau Ffestiniog to Llandudno Junction service, is pictured heading towards Dolwyddelan with a Class 108 unit leading. 25 August 1990.

68. Heading away from the camera, Class 116 DMU set T327, forming the 12.15 Blaenau Ffestiniog to Llandudno service, runs past piles of slate at the old quarries as it heads away from Blaenau Ffestiniog and approaches Blaenau Tunnel. The tunnel, over two miles long, is the longest single-track railway tunnel in Great Britain. 23 June 1990.

69. DMU set T327, forming the 11.00 Llandudno to Blaenau Ffestiniog, passes the slate and disused inclines at the old quarries as it nears the end of its journey. 23 June 1990.

70. A Class 101 DMU, forming the 15.07 Blaenau Ffestiniog to Llandudno, heads away from Blaenau Ffestiniog. On the right, narrow gauge Class NGG16 2-6-2+2-6-2T Beyer Garratt steam locomotives are stored by the Ffestiniog Railway for use on the Welsh Highland Railway. 20 September 1997.

71. RT Railtours' 'The Slate & Narrow' ran from Skipton to Blaenau Ffestiniog. Here, the returning railtour is leaving Blaenau Ffestiniog behind 37211 and 37509. The narrow gauge line of the Ffestiniog Railway is on the right and the coaches of a narrow gauge train can be seen under the bridge. 20 September 1997.

72. Hertfordshire Rail Tours' 'Trawsfynydd Lament' ran from London Euston to Trawsfynydd. Here the railtour is running through Blaenau Ffestiniog behind 56108, out of sight at the front of the train. Brush Type 4 47785 *Fiona Castle* is trailing at the rear. 17 October 1998.

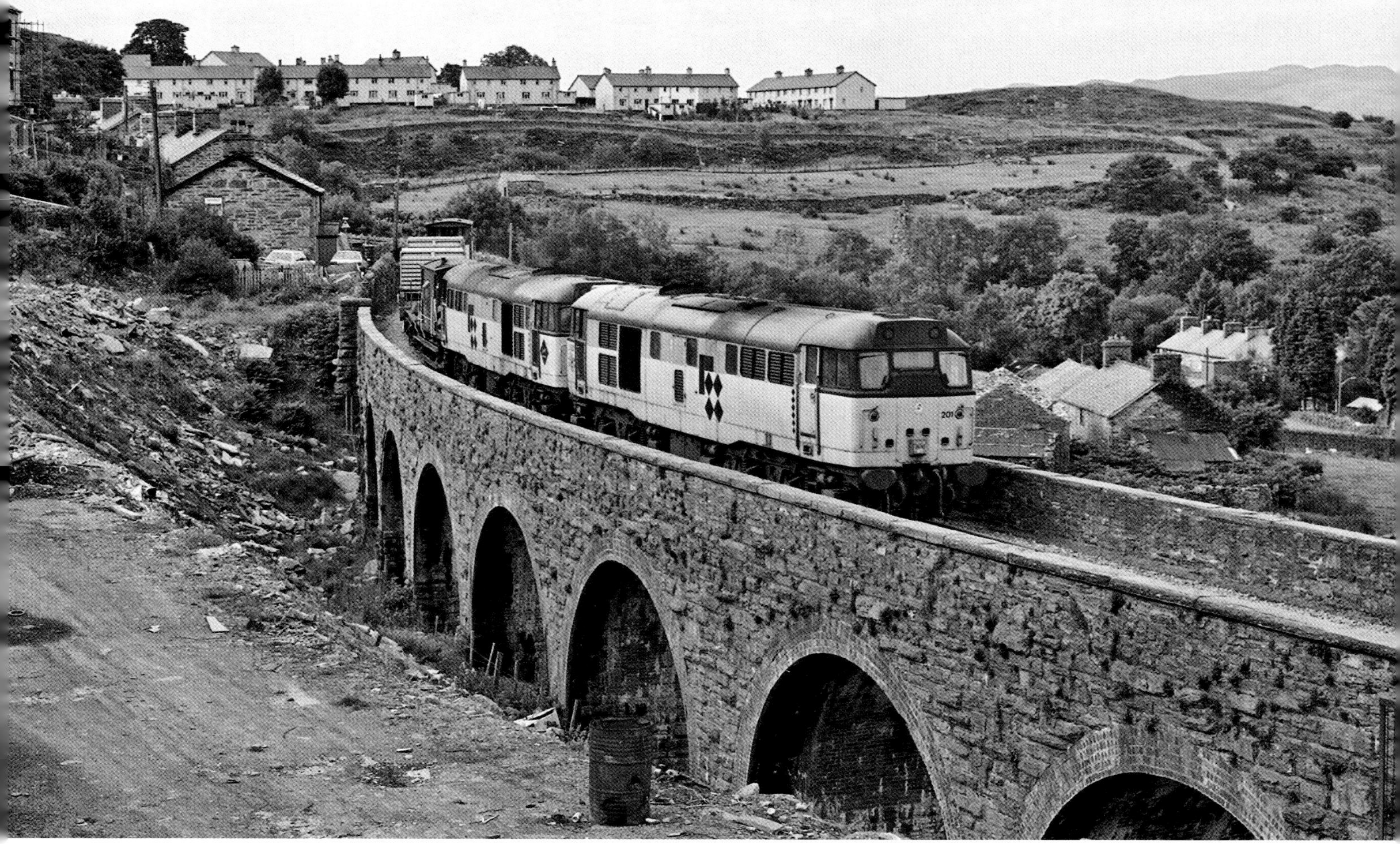

73. The nuclear flask trains from Llandudno Junction were hauled to Blaenau Ffestiniog, where the locomctives ran round and propelled their trains to Trawsfynydd nuclear power station. Here, 31201 and 31199 propel the 7D38 06.30 Llandudno Junction to Trawsfynydd over Gelly Viaduct at Tan-y-Manod, Blaenau Ffestiniog. 29 July 1994. (*Steve Turner Photo*)

74. Brush Type 2 diesels 31201 and 31199 are pictured near Bethania with the 7D39 11.44 Trawsfynydd to Llandudno Junction nuclear flask train. 29 July 1994. (*Steve Turner Photo*)

75. Pathfinder Tours' 'The Trawsfynydd Trekker', from Bristol Temple Meads to Trawsfynydd, passes the closed Maentwrog Road station behind 31207 and 31238. English Electric Type 1s 20075 *Sir William Cooke* and 20187 *Sir Charles Wheatstone*, both in British Rail Telecommunications livery, are at the rear of the train. 10 September 1994.

76. Maentwrog Road station, opened in 1882, was on the Great Western Railway (GWR) line from Bala to Blaenau Ffestiniog. The station was closed to passengers in 1960 and to freight in the following year. Here, 31201 and 31199 pass Maentwrog Road station with the 7D39 11.44 Trawsfynydd to Llandudno Junction. The power station is in the background. 29 July 1994. (*Steve Turner Photo*)

77. Brush Type 2s 31201 and 31199 stand under the loader at Trawsfynydd, waiting to return with the 7D39 11.44 to Llandudno Junction. Regular traffic between Blaenau Ffestiniog and Trawsfynydd finished in 1995 and the last train to carry nuclear material from the power station ran in 1997. 29 July 1994. (*Steve Turner Photo*)

LLANDUDNO JUNCTION TO GAERWEN JUNCTION

78. English Electric Type 3 37421 *The Kingsman* arrives at Llandudno Junction with the 11.22 Bangor to Crewe. The junction of the North Wales Coast line and the line to Llandudno, curving away to the right under the road bridge, can clearly be seen. The power signal box, opened in 1985, is a standard British Railways (London Midland Region) design. 27 July 1996.

79. With Conwy Castle in the background, 31432 arrives at Llandudno Junction with the 10.30 Bangor to Manchester Victoria. The Llandudno line is on the right. 7 August 1993.

80. Leaving the bridge over the River Conwy, Brush Type 4 47613 *North Star* passes Conwy Castle as it heads west with the 1D58 08.50 London Euston to Holyhead. 23 July 1989.

81. Class 37/4 37429 *Eisteddfod Genedlaethol* passes the walls of Conwy Castle as it heads the 14.54 Holyhead to Crewe away from Conwy station. The station platform can be seen through the arch. 29 April 2000.

82. With power car 43065 leading, the InterCity 125 unit, in Virgin Trains livery, forming the 08.25 from London Euston, approaches Conwy station as it heads for Holyhead. This power car was one of eight that were fitted with buffers in 1987 to work as driving van trailers on the East Coast Main Line. 29 April 2000.

83. Class 37/4 37429 *Eisteddfod Genedlaethol* heads the 12.20 Crewe to Holyhead towards Penmaenmawr. Llandudno is in the background, across Conwy Bay. 27 May 2000.

84. With Penmaenmawr in the distance, 31439 *North Yorkshire Moors Railway* heads the 10.30 Bangor to Manchester Victoria along the coast towards Conwy. 11 September 1993.

85. Penmaenmawr station, opened in 1849, is now unstaffed, with the old station building in private use. Here, class 101 DMU set L840, forming a train to Llandudno, pauses at the station. Penmaenmawr signal box, at the east end of the down platform, was opened in 1952. It is a standard London Midland Region design and replaced an earlier box after a collision involving the up 'Irish Mail' and a light engine in 1950. 6 February 1998. (*Paul Dorney Photo*)

86. English Electric Type 3 37408 *Loch Rannoch* stands at the head of a rake of ballast wagons at Penmaenmawr quarry sidings as its train is filled with ballast. 6 February 1998. (*Paul Dorney Photo*)

87. With its train loaded with ballast, 37420 *The Scottish Hosteller* heads out of the quarry sidings at Penmaenmawr. Since it is winter, the North Wales Expressway, in the background, is quiet with little traffic. 23 February 1995. (*Paul Dorney Photo*)

88. Passing the quarry sidings at Penmaenmawr, 37418 *East Lancashire Railway* heads the 1D80 10.15 Blackpool North to Holyhead towards Bangor. 24 June 1995.

89. In Civil Engineers' yellow and grey livery, 37099 *Clydebridge* approaches Penmaenmawr with the 1K67 13.23 Bangor to Crewe. 24 June 1995.

90. The InterCity 125 unit, forming the 08.25 London Euston to Holyhead, crosses Penmaenmawr Viaduct as it heads away from Penmaenmawr. The viaduct is located below the A55 North Wales Expressway. 27 May 2000.

91. Class 101 DMU 101 685, forming an empty stock movement from Bangor to Llandudno Junction, runs alongside Conwy Bay as it heads away from Llanfairfechan. In 1994, 101 685 was repainted in a green livery, previously carried by these units, for use on summer services on the Blaenau Festiniog line. The unit was known as *Daisy* after the DMU in the Reverend Awdry's books. 9 September 1995.

92. Split box Class 37 37107, in Railfreight Distribution livery, heads the Holyhead to Stanlow empty fuel tanks past Llanfairfechan. 24 June 1995.

93. Bangor station was opened in 1848 by the Chester and Holyhead Railway. It is located between two tunnels, Bangor Tunnel to the east and Belmont Tunnel to the west. A branch line to Bethesda, opened in 1884, ran from here. The line closed to passengers in 1951 and to freight in 1962. Here, Class 150 Sprinter 150 108, forming the 19.09 Bangor to Crewe, waits for departure time at Bangor station. 24 July 1989.

94. Viewed from above Bangor Tunnel, an unidentified Class 47 heads east from Bangor station with a train of tanks from Associated Octel at Amlwch. The four tanks at the front of the train contain ethylene dibromide and the white tanks will have been used to transport liquid chlorine to Amlwch. The old locomotive shed building is behind the trees, on the left. 30 March 1991.

95. Leaving Belmont Tunnel, 37422 *Robert F. Fairlie* passes Bangor signal box as it approaches the station with the 1N05 13.10 Holyhead to Blackpool North. Bangor signal box, formerly Bangor No. 2, is a LNWR Type 5 design. It was opened in 1923, replacing an earlier box. 17 September 1994.

96. Llanfair PG or Llanfairpwll railway station, on Anglesey, was opened in 1848. The station is also known by the longer name of Llanfairpwllgwyngyllgogerychwyrndrobwllllantysiliogogogoch, but this is only for tourists. In English the name means 'Saint Mary's Church in the hollow of white hazel near a rapid whirlpool and the Church of Saint Tysilio near the red cave'. 24 July 1989.

97. Class 150 Sprinter 150 127, forming the 15.20 Stalybridge to Holyhead service, approaches Llanfair PG station. Llanfair PG signal box, built circa 1871 to a LNWR design, is located at the Station Road level crossing. The Marquess of Anglesey's Column is on the hill, in the background. 24 July 1989.

98. Class 108 DMU set LO266, forming the 10.50 Holyhead to Crewe, passes Gaerwen signal box, located at the east end of the former station site. The station was closed in 1966. The signal box, formerly Gaerwen No. 1, is a LNWR Type 4 box. Gaerwen is the junction of the branch line to Llangefni and Amlwch, which can be seen curving away to the right behind the train. 30 March 1991.

THE AMLWCH LINE

99. The Anglesey Central Railway, opened in the 1860s, ran from the port of Amlwch, through Llangefni, to Gaerwen Junction. Passenger services were withdrawn in 1964 while freight services continued on the line until 1993. Here, Brush Type 4 47363 stands outside the Associated Octel works at Amlwch with the 7D05 08.55 to Llandudno Junction. The train will be taken forward from there to Ellesmere Port. 16 August 1993. (*Steve Turner Photo*)

100. Brush Type 4 47363 waits for permission to proceed to Associated Octel from the former exchange sidings at the Amlwch station site with the 7D04 05.42 from Llandudno Junction. 16 August 1993. (*Steve Turner Photo*)

101. Brush Type 4 47363 runs through the countryside near Llanerch-y-medd with the 7D05 08.55 Amlwch Associated Octel to Llandudno Junction. On this occasion, the train is made up of one ethylene bromide tank, four empty liquid chlorine tanks, three vans used as barriers and two brake vans. 16 August 1993. (*Steve Turner Photo*)

102. Llangefni station, opened in the 1860s, was closed in 1964. The station building is now a private residence. Here, 47363 passes the old station building with the 7D05 08.55 Amlwch Associated Octel to Llandudno Junction. 16 August 1993. (*Steve Turner Photo*)

GAERWEN JUNCTION TO HOLYHEAD

103. Malltraeth Viaduct, located west of Gaerwen and north of Malltraeth village, carries the North Wales Coast line over the Afon Cefni. Here, 47555 *The Commonwealth Spirit* crosses the nineteen-arch viaduct with the 17.15 Holyhead to London Euston. 30 March 1991.

104. Ty Croes station was opened in 1848 and the signal box, built to a Saxby and Farmer design, in 1872. Here, Class 108 DMU set CH378, forming the 15.28 Llandudno to Holyhead, passes the signal box as it arrives at Ty Croes station. 24 July 1989.

105. In Regional Railways livery, English Electric Type 3 37422 *Robert F. Fairlie* heads the 14.53 Holyhead to Crewe through the open countryside near Rhosneigr. 3 August 1996.

106. One week before the previous photograph, 37422 *Robert F. Fairlie* is heading the 14.53 Holyhead to Crewe away from Valley. Valley's down distant signal is behind the train. 27 July 1996.

107. Former Glasgow Eastfield Class 37/4 37408 *Loch Rannoch*, complete with the West Highland Terrier emblem of Eastfield locomotive depot, heads the 13.53 Holyhead to Crewe near Valley. 31 August 1996.

108. Class 156 Sprinter 156 427, forming the 15.20 Holyhead to Llandudno, arrives at Valley. The station, originally opened in 1849, was closed from 1966 to 1982. 24 July 1989.

109. Sprinter 156 427, forming the 15.20 Holyhead to Llandudno, passes the signal box as it departs from Valley. The box, opened in 1904, was built to a LNWR Type 5 design. 24 July 1989.

110. Class 101 DMU 101 685, forming the 13.18 Llandudno Junction to Holyhead service, is pictured approaching Holyhead. The DMU has been repainted in green livery. 13 August 1994.

111. Hertfordshire Rail Tours' 'The Rat Requiem' ran from London Kings Cross to Holyhead. Sulzer Type 2 D7672 *Tamworth Castle* headed the train from Leeds to Holyhead. Here, the special is approaching Holyhead behind D7672. The large headboard reads 'The Rat Requiem D7672 Final Journey 30-3-1991'. The locomotive depot is to the left of Holyhead signal box, which was built to a 1930s' London Midland and Scottish Railway (LMSR) design. 30 March 1991.

112. English Electric Type 3 37408 *Loch Rannoch* heads the 1K73 15.55 Holyhead to Crewe away from Holyhead. The signal box and the locomotive shed are on the right. The shed, with one through-road, was opened in 1989. A Class 08 shunter is standing near the fuelling point. 13 August 1994.

113. The present-day Holyhead station on Holy Island, Anglesey, was opened by the LNWR in 1866. Trains connect with services to Ireland from Holyhead Ferry Terminal. Sealink ferry *St Columba*, operating between Holyhead and Dun Laoghaire, is at the terminal, in the background. There was a container terminal next to the station which closed in 1991 when the traffic was transferred to Liverpool. The old train shed is on the far right. Here, Class 08 diesel-electric 08737 shunts passenger stock at Holyhead station. A Class 47 is on the left. 24 July 1989.

114. Class 37/4 37418 *East Lancashire Railway* stands with its train at the service point at Holyhead station, while 37408 *Loch Rannoch* waits for departure time, in the platform on the right, with the 1K73 15.55 Holyhead to Crewe. 13 August 1994.

CHESTER TO WREXHAM

115. The railway line between Chester and Shrewsbury was built in 1846 as the Shrewsbury and Chester Railway. Both Chester and Shrewsbury are on the English side of the border, but part of the line runs within Wales. This is the section that we will look at here. Having travelled along the North Wales Coast, the InterCity 125 unit forming the 09.05 Holyhead to London Euston departs from Chester, now without semaphore signals after the completion of resignalling around the station. This train will have already passed Saltney Junction, just in England, where the Holyhead line and the former GWR line to Shrewsbury divide. 25 February 1995.

116. Rossett station, located about six miles north of Wrexham, was opened in 1846. It closed to passengers in 1964 and to freight in 1968. Here, English Electric Type 3 37117 heads north, past Rossett signal box, with a short permanent way train. The signal box, a British Railways (Western Region) Type 37a box with a forty-lever frame, was opened in 1960, replacing two earlier boxes. Just over a week after this photo was taken, the line through Rossett was singled and the signal box was closed. 25 January 1986.

117. Permanent way work is in progress as a Metro-Cammell Class 101 DMU, forming a Chester to Shrewsbury service, passes the old station site at Rossett. 25 January 1986.

118. 'Cromptons' 33065 *Sealion* and 33063 provided the motive power for Hertfordshire Rail Tours' 'The Whistle Test' from London Waterloo to Chester. Here, the special train is seen heading north, near Rossett, on the single line between Wrexham and Saltney Junction. 25 February 1995.

119. A Class 101 DMU, forming a Chester to Shrewsbury service, arrives at Wrexham General. In the background are two signal boxes, the former Great Central Railway (GCR) Wrexham Exchange box on the left, and the ex-GWR Wrexham North on the right. Wrexham North signal box, built in 1883, was a McKenzie & Holland Type 3 design. It was closed in 1986. 25 January 1986.

120. A Class 101 DMU, forming a Shrewsbury to Chester service, approaches Wrexham General station. The station, opened in 1846 and rebuilt in 1912, is on the former GWR route from Shrewsbury to Chester. The second station at Wrexham, Wrexham Central, is the southern terminus of the Wrexham to Bidston line. The current Central station was built in 1998, replacing the original station, opened in 1887. 25 January 1986.

121. A Class 101 DMU, forming a Shrewsbury to Chester service, approaches Croes Newydd North Fork signal box at Wrexham, as it heads away from the camera. Wrexham General station is under the bridge in the background. Croes Newydd North Fork signal box, located at Watery Road level crossing, is a GWR type 27c design. It opened around 1906. 25 January 1986.

WREXHAM TO SHOTTON AND HAWARDEN BRIDGE

122. Having arrived from Wrexham Central, Class 142 Pacer 142 028, forming a Wrexham to Bidston service, waits at Wrexham General platform 4. This part of the station was originally Wrexham Exchange station. The old station building, now demolished, is on the left. 25 January 1986.

123. Pacer 142 028, forming a Wrexham Central to Bidston service, heads away from the camera as it leaves Wrexham General station. Wrexham Exchange signal box, on the former GCR route, is on the left. The signal box, opened in 1912, is built to a GCR Type 5 design. It was moved to Quorn and Woodhouse on the preserved Great Central Railway after it closed in 1988. 25 January 1986.

124. The former GCR line between Wrexham and Bidston was built by the Wrexham, Mold and Connah's Quay Railway and the North Wales and Liverpool Railway. Here, 56056 heads the 6M84 Margam to Dee Marsh steel coils past Gwersyllt, two miles north of Wrexham. 11 February 1994. (*Paul Dorney Photo*)

125. English Electric Type 3 37261 *Caithness* approaches Caergwrle with the 6S72 Dee Marsh to Mossend freight. 14 January 1998. (*Paul Dorney Photo*)

126. British Rail Class 60 60057 *Adam Smith* passes Hope, between Caergwrle and Penyffordd, with the 6M64 17.10 Elgin to Dee Marsh Junction timber train. 16 July 1994. (*Steve Turner Photo*)

127. British Rail Class 56 56073 *Tremorfa Steel Works* passes Penyffordd with the 6V78 15.30 Dee Marsh Junction to Margam Yard steel empties. Penyffordd signal box is a British Railways (London Midland Region) type 15 design, built in 1972. 16 July 1994. (*Steve Turner Photo*)

128. Shotton station is on two levels. Shotton High Level is on the Wrexham to Bidston line, while Shotton Low Level is on the North Wales Coast line. Here, 60082 *Mam Tor* passes Shotton High Level station with the 6F46 17.30 Dee Marsh Junction to Arpley Sidings empty timber train. 23 July 1994. (*Steve Turner Photo*)

129. Class 142 Pacer 142 048, forming the 2J75 14.32 Bidston to Wrexham Central, leaves Hawarden Bridge station and approaches the bridge over the River Dee. The station was opened by the London and North Eastern Railway (LNER) in 1924 as Hawarden Bridge Halt and was renamed Hawarden Bridge in 1954. 3 September 1994. (*Steve Turner Photo*)

WREXHAM TO CHIRK

130. With Croes Newydd North Fork signal box in the background, Class 56 56071 heads south from Wrexham with the 6J70 Warrington Arpley to Chirk Kronospan timber train. The waste ground on the left is the site of the former Croes Newydd locomotive depot. 25 February 1998. (*Paul Dorney Photo*)

131. Class 37/9 37902, fitted with a Mirrlees MB275T engine, passes Bersham Colliery, near Rhostyllen, as it heads south towards Ruabon. The connection with the main line was severed after the colliery closed in December 1986. 30 May 1987.

132. The boarded-up Ruabon signal box appears to be near to the end of its life, switched out and with its nameplate removed. Here, Class 150 Sprinter 150 123, forming a Shrewsbury to Chester service, passes the box as it approaches Ruabon station. The box is the former GWR Ruabon Middle signal box. 13 June 1987.

133. Ruabon station opened in 1846 on the Shrewsbury to Chester line, later part of the GWR route from London Paddington to Birkenhead. It was also the terminus of the Ruabon to Barmouth line which was opened in stages between 1861 and 1867, finally closing in 1968. Here, a Class 120 DMU, forming a Chester to Shrewsbury service, waits to depart from the station. 1 March 1986.

134. Class 150 Sprinter 150 129, forming a Chester to Wolverhampton service, passes Ruabon's down distant signal as it heads south towards Chirk. The finials were removed from the GWR signals after the Shrewsbury to Chester line was transferred from the Western to the London Midland Region in 1963. 30 May 1987.

135. Built in 1848, the nineteen-arch Cefn Mawr Viaduct crosses the River Dee, south of Ruabon. Here, a DMU, forming a Chester to Wolverhampton service, heads south over the viaduct. 30 May 1987.

136. English Electric Type 3 37380, in Mainline livery, stands at the Kronospan Works at Chirk after arriving with the 6J70 from Arpley Yard, Warrington. Kronospan, a manufacturer of wood-based panel products, is located just north of Chirk station. 13 May 1998. (*Paul Dorney Photo*)

137. A Class 150 Sprinter, forming the 07.53 Wolverhampton to Crewe, crosses Chirk Viaduct over the River Ceiriog. Chirk Aqueduct, carrying the Llangollen Canal, is on the left. 4 March 1989.

138. Built in 1848, Chirk Viaduct carries the Shrewsbury to Chester railway across the Ceiriog Valley. The viaduct is approximately 849ft long with sixteen arches. In front of the railway viaduct is Chirk Aqueduct, built in 1801. It is 710ft long with ten arches and carries the Llangollen Canal across the valley. Here, a DMU heads north across the viaduct. 26 April 1975.

LLANGOLLEN RAILWAY

139. The Llangollen Railway is a standard gauge heritage railway which runs for ten miles between Llangollen and Corwen along the former GWR route from Ruabon to Barmouth Junction, closed in 1965. The railway reopened a short section of track for tourists in 1975 and was extended to Berwyn in 1986, reaching Carrog in 1996. Services to Corwen commenced in 2014. Although most trains on the railway are steam-hauled, diesel traction is also used. Here a two-car Class 104 DMU, made up of M50528 and M50454, stands in Llangollen station. 22 October 1994.

140. Berwyn station, originally opened in 1865, is the first station stop for trains departing from Llangollen. Sulzer Type 2 24081 sets out across Berwyn Viaduct alongside the River Dee with a train from Llangollen. The narrow gauge Bala Lake Railway is also built on the trackbed of the former Ruabon to Barmouth line, along the shore of Bala Lake. 22 October 1994.

WELSHPOOL & LLANFAIR LIGHT RAILWAY

141. The Welshpool & Llanfair Light Railway is a 2ft 6in gauge narrow gauge railway running 8½ miles from Welshpool to Llanfair Caereinion. The railway was closed by British Railways in 1956 and reopened as a heritage railway in 1963. Here, one of the original locomotives, 1 *The Earl* of 1902, is pictured at Llanfair Caereinion. 26 May 1969.

142. Opened in 1903, the Welshpool & Llanfair Light Railway originally ran from a station next to the standard gauge line in Welshpool, but now starts from a station at Raven Square on the edge of the town. Two locomotives were built for the line by Beyer Peacock in 1902 and were stored at Oswestry Works after the railway closed. Here, the second locomotive, 2 *Countess*, is pictured at Llanfair Caereinion, eleven years after the previous photo of *The Earl* at the same location. 25 August 1980.

THE CAMBRIAN FROM WELSHPOOL TO PWLLHELI

143. The railway line from Welshpool to Aberystwyth and Pwllheli was originally part of the Cambrian Railway system. The former Cambrian Railways narrow gauge line to Llanfair Caereinion, closed by British Railways in 1956 and now in private ownership, also started at Welshpool. Here, a six-car DMU, forming the 08.40 Pwllheli to Birmingham New Street service, departs from Welshpool. This splendid station, built by the Oswestry and Newtown Railway, was opened in 1860. 25 July 1985.

144. Class 37/4 37431 departs from Welshpool with the up 'Cambrian Coast Express', the 07.22 Aberystwyth to London Euston. Note the steam era-style 'Cambrian Coast Express' headboard, as carried by the named express from London Paddington to Aberystwyth and Pwllheli in the 1950s and early 1960s. In 1992, the railway at Welshpool was realigned to make room for the A483 Welshpool bypass and the old station building is now privately-owned, located on the other side of the road. The new station is an island platform. 26 July 1986.

145. The signalman waits to take the token from the crew of 37429, as it departs from Welshpool with the up 'Cambrian Coast Express' from Aberystwyth to London Euston. The token for the next section is ready for collection on the token exchange equipment, in front of the signal box. The signal box was built in 1897 to a LNWR design. 19 July 1986.

146. Sulzer Type 2 25037 stands at Welshpool after failing with the Stanlow to Aberystwyth oil tanks. 30 August 1986.

147. Carrying the 'Cambrian Coast Express' headboard on its nose, 37429 *Sir Dyfed/County of Dyfed* passes Forden level crossing with the 15.40 London Euston to Aberystwyth. Forden station closed in 1965, but the station house and signal box were adapted for private use. The Dutton Type 3 signal box was built in 1897 for the Cambrian Railways. 25 July 1987.

148. Abermule was the junction for the short branch to Kerry, closed in 1956. Here, 37426 *Y Lein Fach/Vale of Rheidol* passes with the up 'Cambrian Coast Express', the 07.22 Aberystwyth to London Euston. Note the different style of headboard on the locomotive's nose. The Dutton & Co Type 1 signal box was opened in 1891 and the station, opened in 1860, was closed in 1965. One of the platforms still remains, on the right. Two trains collided head on between Abermule and Newtown in January 1921, resulting in the deaths of seventeen people. 21 June 1986.

149. Newtown station, opened in 1863, is located to the east of Moat Lane Junction, where the line to Rhayader and Three Cocks Junction diverged from the Cambrian line to Machynlleth. Here, 37506 leads the 1J20 07.30 London Euston to Pwllheli away from Newtown. This train was normally double-headed and the second locomotive is 37050 on this occasion. 26 July 1986.

150. A second departure from Newtown on the same day, this time behind Class 37/4 37430 *Cwmbran*, fitted with electric train supply. The working is the 1J24 09.35 London Euston to Aberystwyth. The large building in the background, dating from 1895, is the Royal Welsh Warehouse, the former premises of the Pryce Jones textile company. 26 July 1986.

151. Newtown signal box, with its associated token exchange equipment, stands at the southwest end of the station. This is the former Newtown South box, a Dutton & Co Type 3 design built in the 1890s. Dutton & Co Ltd of Worcester undertook most of the signalling work on the Cambrian Railway at the time. 17 September 1988.

152. A Metro-Cammell Class 101 DMU is stabled at Newtown, with a second unit behind. 5 July 1986.

153. Class 150 Sprinter 150 119, forming the 17.05 Shrewsbury to Aberystwyth service, passes the signal box as it departs from Newtown. Note the mixture of upper and lower quadrant signals. 21 June 1986.

154. Caersws station, built by the Newtown and Machynlleth Railway in the 1890s and once the junction for the Van Railway, is located to the west of Moat Lane Junction. Here, passengers wait at the station as 37431, as yet unnamed, arrives with the 1J24 09.35 London Euston to Aberystwyth. 9 July 1986.

155. English Electric Type 3 37431 *Sir Powys/County of Powys* waits at Caersws with the 07.13 Aberystwyth to London Euston. The passing loop has been removed and 37431 has been named in the three years since the previous photograph. The signal box, on the end of the platform, is a Dutton Type 1 design, dating from 1892. 9 September 1989.

156. English Electric Type 3 37431 departs from Caersws with the 09.35 London Euston to Aberystwyth. 19 July 1986.

157. The GWR Type 25 signal box at Carno was moved from Newton Abbot around 1928 to replace an earlier box. The signal box closed in 1988 when the level crossing was modernised. Here, 37428 passes Carno station, closed in 1965, with the 1J24 09.35 London Euston to Aberystwyth. 27 September 1986.

158. Talerddig railway station, between Carno and Llanbrynmair, was opened in 1900 and closed in 1965, the station building being demolished soon after closure. The signal box is a McKenzie and Holland design, built in 1874. Here, Sulzer Type 2 25211 waits for the signalman to bring the token before proceeding east to assist a failed locomotive. 14 June 1986.

159. Signalling equipment built by Tyer & Co Ltd was used in Talerddig signal box. Tyer & Co of Ashwin Street, Dalston, founded in 1850, were manufacturers of electrical and mechanical railway signal equipment. 17 September 1988.

160. The token for the single line to Caersws is ready for collection as 37428 passes Talerddig signal box with the 1A81 15.40 Aberystwyth to London Euston. 13 September 1986.

161. The 15.30 Pwllheli to Birmingham New Street climbs the 1 in 56 gradient towards Talerddig, at Pont Bell, behind Class 31/1 locomotives 31147 and 31146 in Civil Engineers' yellow and grey livery. 25 July 1992.

162. Llanbrynmair station, between Cemmaes Road and Talerddig, was closed in 1965, one of many stations to be closed on the Cambrian line as a result of the Beeching cuts. Here, 37431 leads the 1J18 07.25 Birmingham New Street to Aberystwyth past the old station building at Llanbrynmair. 27 September 1986.

163. Pictured west of Llanbrynmair, 37298 and 37158 head the 1J19 07.40 London Euston to Pwllheli towards Cemmaes Road. 21 May 1988.

164. The 1A50 11.10 Aberystwyth to London Euston rounds the curve at Commins Coch, between Cemmaes Road and Llanbrynmair, behind 37427 *Bont Y Bermo*. The A470 road is in the background. 30 August 1986.

165. A Metro-Cammell Class 101 DMU, forming the 13.35 Aberystwyth to Shrewsbury service, passes the old Cemmaes Road station building. The station was closed in 1965. 14 June 1986.

166. In Railfreight livery, 37501 approaches Cemmaes Road with the 1A50 11.10 Aberystwyth to London Euston. The signal box, a GWR Type 28b design, was closed in 1984. Cemmaes Road was the junction for the line to Dinas Mawddwy, closed in 1952. 21 June 1986.

167. With the crew preparing to hand over the token for the single line, 37428 and 37430 *Cwmbran* arrive at Machynlleth with the 1J20 07.30 London Euston to Pwllheli. 14 June 1986.

168. After participating in the events around the reopening of Barmouth Bridge to locomotive hauled trains, Class 150 Sprinter 150 122 prepares to head back from Machynlleth to Shrewsbury. A second Sprinter, 150 116, and Class 37/4 37428 can be seen in the locomotive depot in the background. The signal box was a British Railways (Western Region) Type 37a design, opened in 1960. 13 April 1986.

169. Machynlleth station was built by the Newtown and Machynlleth Railway. Opened in 1863, it was absorbed into the Cambrian Railways in 1868. Here, viewed from the station footbridge, 37427 *Bont Y Bermo* arrives at Machynlleth with the 1A50 11.10 Aberystwyth to London Euston. 23 August 1986.

170. A small engine shed opened in 1863 at Machynlleth and was later expanded by the Cambrian Railways. Present on the shed are 37430 *Cwmbran,* a DMU with its engine running, and Ivatt Class 2 2-6-0 46443. The steam locomotive, based on the Severn Valley Railway, was one of the locomotives used on excursion trains along the Cambrian Coast line in 1987. 29 August 1987.

171. Sulzer Type 2 25191 stands at Machynlleth locomotive depot. The Class 25s were used on permanent way duties in the area at this time. 19 July 1986.

172. Dovey Junction railway station is the junction of the lines to Machynlleth, Aberystwyth and Pwllheli. Opened in 1863 as Glandovey Junction, the station was renamed Dovey Junction in 1904. Here, 37177 and 37427 *Bont Y Bermo* start the 1A81 15.40 Aberystwyth to London Euston away from the Dovey Junction station stop. The signal box, visible on the station platform, was a British Railways (Western Region) Type 37a. It was closed in 1988. Dovey Junction is one of the few stations in Great Britain with no road access. 14 June 1986.

173. The 1A81 15.40 Aberystwyth to London Euston approaches Dovey Junction station behind 37426 *Y Lein Fach/Vale of Rheidol* and 37428. 5 July 1986.

174. With Smugglers' Cove Boatyard in the foreground, 37431 *Sir Powys/County of Powys* heads the 15.05 Pwllheli to London Euston past the site of Abertafol halt, between Penhelig and Dovey Junction. 9 September 1989.

175. Pictured running alongside the River Dovey estuary near Abertafol, 37257 and 37431 *Sir Powys/County of Powys* head the 07.40 London Euston to Pwllheli towards Penhelig. 29 August 1987.

176. With the River Dovey estuary on the right, 37198 and 37142 head the 07.40 London Euston to Pwllheli towards Penhelig station. 1 October 1988.

177. The 1A85 17.08 Porthmadog to London Euston heads away from Penhelig behind 37257 and 37293. This train left Pwllheli at 15.30 as the 2J26 Pwllheli to Porthmadog, where it waited for more than an hour before departing for London as the 1A85. 13 September 1986.

178. Rounding the curve after leaving the Aberdovey station stop, 37429 *Eisteddfod Genedlaethol* heads the 1A46 09.40 Pwllheli to London Euston towards Penhelig. 9 September 1989.

179. Towyn station, opened in 1863, was renamed Tywyn in 1975. Tywyn still has two platforms and a passing loop and here 37430 *Cwmbran* exits the loop as it departs from Tywyn with the 08.00 Pwllheli to London Euston, 'The Snowdonian'. 8 August 1987.

180. Tywyn signal box is a GWR Type 25 design, dating from the 1890s. The box was moved from Maidenhead around 1923 and replaced two earlier signal boxes. Here, tokens are exchanged at the box as 37430 *Cwmbran* and 37248 arrive with the 1J20 07.30 London Euston to Pwllheli. 5 July 1986.

181. A Class 150 Sprinter, forming the 09.10 Dovey Junction to Pwllheli, passes a camping and caravan site next to the coast, south of Llangelynin. 25 July 1987.

182. With Fairbourne and Barmouth in the distance, 37430 *Cwmbran* heads the 08.00 Pwllheli to London Euston, 'The Snowdonian', towards Llwyngwril. 25 July 1987.

183. English Electric Type 3s 37427 *Bont Y Bermo* and 37240 run along the Friog Cliffs above Barmouth Bay, between Fairbourne and Llwyngwril, with the 1A84 15.30 Pwllheli to London Euston. This is the site of an accident in 1933, when a train collided with a landslip. 25 July 1987.

184. Barmouth Bridge is a wooden railway viaduct, approximately 2300ft long, across the River Mawddach estuary. Here, with the sun low in the west, 37431 *Sir Powys/County of Powys* sets out across the bridge with the 16.56 Pwllheli to London Euston. 14 July 1990.

185. Following damage to Barmouth Bridge from marine woodworm, the bridge was closed to locomotive-hauled trains in 1980. After repairs, locomotives were again allowed to cross the bridge in 1986 and a special train, the 'Barmouth Bridge Special', organised by Hertfordshire Rail Tours and F & W Tours ran from London Euston to Barmouth in the April of that year. Here, the train is seen crossing the bridge and approaching Barmouth behind 37426 and 37427. 13 April 1986.

186. English Electric Type 3 37427 was officially named *Bont y Bermo* at Barmouth after arriving with the 'Barmouth Bridge Special' on 13 April 1986. Bont Y Bermo means Barmouth Bridge. This photograph of the nameplate and the plate showing the Barmouth Bridge was taken at Newtown while 37427 was heading the up 'Cambrian Coast Express'. 25 May 1987.

187. After leaving Barmouth station, 37431 *Sir Powys/County of Powys* heads for Barmouth Bridge with the 16.56 Pwllheli to London Euston. 23 June 1990.

188. Class 150 Sprinter 150 136, forming the 14.50 Shrewsbury to Pwllheli service, approaches Barmouth station. Barmouth South signal box, on the right, is a Dutton & Co Type 1 design built in 1890. The box has since been moved to Glyndyfrdwy station on the Llangollen Railway. The semaphore signals are a mixture of upper and lower quadrant designs. 7 June 1986.

189. Barmouth station was opened in 1867. The main part of the station is located beyond the signal box on the other side of the level crossing, where a Class 150 Sprinter waits in the down platform. Here, 37430 *Cwmbran* and 37506 depart from the station with the 1A85 17.08 Porthmadog to London Euston, past the old Dolgellau line platform. The platform became redundant when the branch from Barmouth Junction, later named Morfa Mawddach station, to Dolgellau closed in 1965. 30 August 1986.

190. English Electric Type 4 40122/D200 and Sulzer Type 2 25288 head north from Barmouth station with Traintours' 'The Welsh Thunderer' from Stalybridge to Pwllheli. These locomotives worked from Stalybridge to Pwllheli, with 37426 *Y Lein Fach/Vale of Rheidol* and 37428 returning the train to Shrewsbury. From there, 40122 took the train back to Stalybridge. 7 June 1986.

191. Sprinter 150 127, forming the 14.54 Pwllheli to Machynlleth service, crosses the trestle bridge over the Afon Artro at Pensarn Harbour, close to Pensarn station, between Llandanwg and Llanbedr. 14 July 1990.

192. F & W Tours' 'Snowdonian II' railtour ran from London Paddington to Pwllheli. Class 50 50019 *Ramillies* headed the train from Paddington to Bristol Temple Meads, with 47501 taking charge of the train as far as Wolverhampton via Gloucester. Sulzer Type 2 locomotives 25201 and 25037 then took the train on to Pwllheli. The special train is seen here arriving at Harlech, en route to Pwllheli. 24 August 1986.

193. A Metro-Cammell Class 101 DMU, forming a Machynlleth to Pwllheli service, approaches Harlech station. The signal box is a GWR Type 8a, built in the early 1920s. 6 September 1986.

194. Viewed from Harlech Castle, a DMU forming the 11.25 Pwllheli to Machynlleth waits at Harlech station for a northbound service to cross. 6 September 1986.

195. With the castle dominating the scene, a Metro-Cammell Class 101 DMU, forming a Pwllheli to Machynlleth service, waits for the road at Harlech before continuing its journey south. 6 September 1986.

196. A toll road and railway bridge crossed the Afon Dwyryd between Penrhyndeudraeth and Llandecwyn. Here, 37427 *Bont Y Bermo* and 37429 *Eisteddfod Genedlaethol* head the 1A84 15.30 Pwllheli to London Euston across the bridge towards Llandecwyn. After deterioration of the timber structure, the bridge was replaced in 2014. 8 August 1987.

197. After crossing the bridge over the Afon Dwyryd, 37431 *Sir Powys/County of Powys* catches up with a Fiat Uno on the toll road, as it heads for Penrhyndeudraeth station with the 09.03 London Euston to Pwllheli, 'The Snowdonian'. 23 June 1990.

198. Located between Minffordd and Llandecwyn, Penrhyndeudraeth station was opened in 1867. Here, Class 150 Sprinter 150 138, forming a Pwllheli to Machynlleth service, is pictured departing from the station. 7 June 1986.

199. Portmadoc station was opened in 1867 and renamed Porthmadog in 1975. The Dutton Type 2 signal box, built in the 1890s, was extended by the GWR in 1932. Here, 37431 and 37430 *Cwmbran* depart from Porthmadog with the 1A85 17.08 Porthmadog to London Euston. They had previously worked to Porthmadog on the 2J26 15.30 from Pwllheli. 6 September 1986.

200. The driver of 37427 *Bont Y Bermo* looks back down the train prior to departure from Porthmadog with the 1J20 07.30 London Euston to Pwllheli. The second locomotive is 37428. 19 July 1986.

201. F & W Tours' 'Snowdonian II' railtour waits at Porthmadog during the return journey from Pwllheli to London Paddington. The locomotives are 25037 and 25201. 24 August 1986.

202. With Criccieth Castle in the background, 25037 and 25201 head the returning F & W Tours' 'Snowdonian II' railtour from Pwllheli to London Paddington towards Porthmadog. 24 August 1986.

203. Running next to Criccieth beach and with the castle in the background, 37428 and 37427 *Bont Y Bermo* head the 2J26 15.30 Pwllheli to Porthmadog away from Criccieth. The train will later form the 1A85 17.08 Porthmadog to London Euston. 19 July 1986.

204. Criccieth station was opened by the Aberystwith and Welsh Coast Railway in 1867. Originally with two platforms, the former down platform was made redundant when the passing loop was taken out of use in 1977. Here, Class 150 Sprinter 150 116 arrives at Criccieth station with a special working for the Cambrian Coast Line Action Group from Pwllheli to Barmouth. Note the 'CCLAG' headboard. Between Criccieth and Pwllheli, a London and North Western Railway line to Caernarvon left the Cambrian line at Afon Wen. 24 August 1986.

205. Pwllheli West signal box was a Dutton Type 2 box, opened in 1909. The signal box and lever frame were manufactured by McKenzie & Holland and were moved from Elan Valley Railway Junction. After rationalisation at Pwllheli in the 1970s, the signal box was retained as a ground frame. Here, with the box in the foreground, 37427 *Bont Y Bermo* and 37428 shunt the stock after arriving with the 07.30 from London Euston. 19 July 1986.

206. The terminus of the Cambrian Coast line is at Pwllheli. The station, the second station to be built in the town, was opened near the town centre in 1909. It had an island platform with two tracks, one side of which was abandoned in 1977. Here, a Metro-Cammell Class 101 DMU, forming the 17.28 to Machynlleth, stands in the remaining platform. 7 June 1986.

DOVEY JUNCTION TO ABERYSTWYTH

207. Dovey Junction to Aberystwyth is just over sixteen miles by rail. Here, Class 150 Sprinter 150 137, forming the 14.50 Shrewsbury to Aberystwyth service, approaches the bracket signal to the east of the station as it arrives at Dovey Junction. 5 July 1986.

208. With Dovey Junction in the distance, 37429 *Eisteddfod Genedlaethol* approaches Glandyfi with the 1J24 09.40 London Euston to Aberystwyth. The Cambrian Coast line bridge over the River Dovey can be seen above the locomotive. 29 August 1987.

209. Glandyfi station, opened between Dovey Junction and Borth in the 1860s, was closed in 1965. Here, Class 150 Sprinter 150 137, forming a Machynlleth to Aberystwyth service, passes the old station building. The building was later converted into a private house. 9 August 1987.

210. Sulzer Type 2 25288 heads a short westbound permanent way train, consisting of two ballast wagons, past the former Glandyfi station building. 23 August 1986.

211. Running alongside the River Dovey, 37428 *David Lloyd George* and 37426 *Y Lein Fach/Vale of Rheidol* approach Glandyfi with the 1A44 10.13 Aberystwyth to London Euston. 9 September 1989.

212. English Electric Type 3 diesel-electrics 37680 and 37684 cross the Afon Clettwr, between Glandyfi and Ynyslas, with the 1A44 10.10 Aberystwyth to London Euston. 1 October 1988.

213. En route to Aberystwyth, Class 31/1 31145 heads the Radio 2 Railshow across the Afon Leri at Ynyslas, north of Borth. Ken Bruce presented a programme from the Radio 2 Railshow from Aberystwyth station at 09.30 the following day. 9 August 1987.

214. Two years later, the telegraph poles and wires have gone following the introduction of radio signalling, controlled from Machynlleth, in October 1988. Here, 37426 *Y Lein Fach/Vale of Rheidol* and 37428 *David Lloyd George* head the 1J18 06.02 Birmingham New Street to Aberystwyth across the Afon Leri at Ynyslas. 9 September 1989.

215. Aberystwyth's former standard gauge locomotive shed was a sub-shed of Machynlleth. After closure, it was adapted for use by the narrow gauge Vale of Rheidol Railway. Here, with narrow gauge stock visible outside the shed, 37427 *Bont Y Bermo* departs from Aberystwyth with the 1A33 15.20 Aberystwyth to London Euston. 9 August 1987.

216. Preparing to run round its train after arriving at Aberystwyth, 31145 shunts the Radio 2 Railshow out of the station. After a programme from Aberystwyth the following day, Ken Bruce presented a programme from Llandudno on 11 August, followed by programmes from Morecambe, Scarborough and Skegness. 9 August 1987.

217. English Electric Type 1 diesel-electrics 20133 and 20034 stand at Aberystwyth after arriving with Pathfinder Tours' 'The Rheidol Rambler', the 1Z26 from Bristol Temple Meads to Aberystwyth. Class 31/1 31145 stands alongside, in the platform, with the Radio 2 Railshow train. 9 August 1987.

218. Aberystwyth railway station was opened in 1863 by the Aberystwith and Welsh Coast Railway and extended by the GWR in 1925. It is the terminus of both the Cambrian line, just over 81 miles from Shrewsbury, and the narrow gauge Vale of Rheidol Railway. Here, Class 150 Sprinter 150 142 stands in the station between duties. 14 September 1986.

THE VALE OF RHEIDOL RAILWAY

219. The Vale of Rheidol Railway is a steam-operated 1ft 11 ¾in gauge narrow gauge railway which runs for 11¾ miles from Aberystwyth to Devil's Bridge. After closure, Aberystwyth's standard gauge locomotive depot was converted for use by the Vale of Rheidol Railway. Here, the railway's three 2-6-2T locomotives can be seen at the depot. In the foreground is 8 *Llywelyn* with its boiler removed, 9 *Prince of Wales* is behind, and 7 *Owain Glyndŵr* is outside. 7 May 1984.

220. From the late 1960s until the early 1980s, the Vale of Rheidol locomotives and rolling stock were painted in British Rail's corporate blue livery. Here *Owain Glyndŵr,* in blue livery with the double-arrow painted on the tanks, is stopped in the woods at the top end of the line, high above the Afon Rheidol, while the train crew extinguish a lineside fire started by the hard-working locomotive. 7 April 1969.

221. In the 1980s, the locomotives were repainted into historic liveries. Here, in British Railways lined green livery, *Owain Glyndŵr* blasts through the rock cutting below Devil's Bridge as it nears the end of its journey up the valley. 20 April 1987.

222. Soon after leaving Aberystwyth, 7 *Owain Glyndŵr* crosses the Afon Rheidol, near Llanbadarn station, with a train to Devil's Bridge. *Owain Glyndŵr*, built by the GWR at Swindon Works in 1923, was painted in British Railways lined green livery at this time. The railway was sold into private ownership as an operating railway in 1989. 30 August 1988.

223. After taking water at Nantyronen, 9 *Prince of Wales* resumes its journey to Devil's Bridge. The locomotive carries a yellow ochre livery, similar to that initially applied to the original Davies and Metcalfe locomotives of 1902. The headboard reads 'The Welsh Dragon'. 14 September 1986.

STEAM-HAULED SPECIAL TRAINS

224. Steam specials ran from time to time on the railway lines of North and Mid Wales. In November 1993, Pathfinder Tours' 'The Cymru Coaster' ran from Swindon to Holyhead, with 45596 *Bahamas* taking the train from Crewe to Holyhead and back. Here, the Stanier Jubilee Class 4-6-0, built in 1935, makes a splendid sight as it heads the train through the middle road at Rhyl, in some fine winter sunshine. 27 November 1993.

225. West Country Class 4-6-2 34027 *Taw Valley* departs from Llandudno with the 'North Wales Coast Express' from Crewe to Llandudno and Holyhead. These special trains, organised by the InterCity Charter Train Unit, ran three days a week, using various steam locomotives. *Taw Valley*, originally designed by Oliver Bulleid, was built in 1946. It was rebuilt by British Railways in 1957 to the design of Ronald Guy Jarvis. 23 July 1989.

226. After crossing the River Conwy, West Country Class 4-6-2 34027 *Taw Valley* passes Conwy Castle as it heads for Holyhead with the 'North Wales Coast Express' from Crewe. 23 July 1989.

227. West Country Class 4-6-2 34027 *Taw Valley* heads the 'North Wales Coast Express' past Penmaenmawr on the return journey from Holyhead to Crewe. 23 July 1989.

228. Jubilee Class 4-6-0 45596 *Bahamas* approaches Penmaenmawr with the 'The Cymru Coaster' from Swindon to Holyhead. Llandudno is in the background. 27 November 1993.

229. British Rail's 'Cardigan Bay Express' made two return trips from Machynlleth to Barmouth in May 1987. Here, the morning train from Machynlleth approaches Barmouth, after crossing Barmouth Bridge, behind Manor Class 4-6-0 7819 *Hinton Manor*. Designed by Charles Benjamin Collett, the locomotive was built in 1939 at Swindon Works. 25 May 1987.

230. After working two return trips to Barmouth, Collett 4-6-0 7819 *Hinton Manor* stands next to Machynlleth locomotive depot with the stock of its train. 25 May 1987.

231. Collett 4-6-0 7819 *Hinton Manor,* in BR lined black livery, passes the old station building at Glandyfi, as it heads the 'Cardigan Bay Express' from Machynlleth to Aberystwyth. The train then visited Barmouth before returning to Machynlleth for a second trip to Aberystwyth. 16 August 1987.

BIBLIOGRAPHY

Baker, S. K., *Rail Atlas Great Britain and Ireland*, Haynes Publishing Group, 1980 and 1988
Baughan, P. E., *Regional History of the Railways, North and Mid Wales*, David and Charles, 1980
Boyd, J. I. C., *Narrow Gauge Railways in Mid-Wales*, The Oakwood Press, 1965
British Rail, *British Rail Passenger Timetable(s)*, May 1984-October 1993, British Railways Board, 1984-1993
Christiansen, R., *Forgotten Railways North and Mid Wales*, David and Charles, 1976
Gradients of the British Main Line Railways, Ian Allan Publishing Ltd, 2016
Hillmer, J. & Shannon, P., *Diesels in North and Mid Wales*, Oxford Publishing Co., 1990
Jowett, A., *Jowett's Railway Atlas of Great Britain and Ireland*, Patrick Stephens Ltd, 1989
London Midland Region Track Diagrams, Quail Map Co., 1990
Main Line Railways of North Wales, www. 2D53.co.uk
Marsden, C., *35 Years of Main Line Diesel Traction*, Oxford Publishing Co., 1982
Signalling Study Group, *The Signal Box, A Pictorial History*, Oxford Publishing Co., 1986
Wood, R., *British Rail Locomotives*, Ian Allan Ltd, 1986

INDEX

Aberdovey, 178
Abergele and Pensarn, 39-45
Abermule, 148
Abertafol, 174-175
Aberystwyth, 215-218
Afon Clettwr, 212
Amlwch, 99-100

Bagillt, 10-13
Bangor, 93-95
Barmouth, 184-190, 229
Bersham Colliery, 131
Bethania, 74
Bettws-y-Coed, 65
Blaenau Festiniog, 68-73

Caergwrle, 125
Caersws, 154-156
Carno, 157
Cefn Mawr Viaduct, 135
Cemmaes Road, 165-166
Chester, 2, 115
Chirk, 136-138
Colwyn Bay, 46
Commins Coch, 164
Conwy, 80-82, 226
Criccieth, 202-204

Deganwy, 54-55
Dolwyddelan, 66
Dovey Junction, 172-173, 207

Fairbourne, 183
Ffynnongroyw, 25
Flint, 9
Forden, 147

Gaerwen, 98
Glan Conwy, 62
Glandyfi, 208-211, 231

Gronant, 28
Gwersyllt, 124

Harlech, 192-195
Hawarden Bridge, 129
Holyhead, 110-114
Holywell Junction, 13-17
Hope, 126

Llanbrynmair, 162-163
Llandudno Junction, 48-53, 61, 78-79
Llandudno, 56-60, 225
Llanerch-y-medd, 101
Llanfair PG, 96-97
Llanfairfechan, 1, 91-92
Llangefni, 102
Llangelynin, 181
Llangollen Railway, 139-140
Llwyngwril, 182

Machynlleth, 167-171, 230
Maentwrog Road, 75-76
Malltraeth, 103
Mochdre and Pabo, 47
Mold Junction, 5-8
Mostyn, 18-24

Newtown, 149-153
North Llanrwst, 63-64

Penhelig, 176-177
Penmaenmawr, 83-90, 227-228
Penrhyndeudraeth, 196-198
Pensarn, 191
Penyffordd, 127
Pont Bell, 161
Porthmadog, 199-201
Prestatyn, 29-31
Pwllheli, 205-206

Rhosneigr, 105
Rhyl, 32-38, 224
Roman Bridge, 67
Rossett, 116-118
Ruabon, 132-134

Sandycroft, 3-4
Shotton, 128

Talacre, 26-27
Talerddig, 158-160
Trawsfynydd, 77

Ty Croes, 104
Tywyn, 179-180

Vale of Rheidol Railway, 219-223
Valley, 106-109

Welshpool & Llanfair
   Light Railway, 141-142
Welshpool, 143-146
Wrexham, 119-123, 130

Ynyslas, 213-214